The ChatGPT Millionaire Handbook

Make Money Online With the Power of AI Technology

TJ Books

Contents

Chapter 1

Introduction

In the last few years, Artificial Intelligence has made amazing progress and changed the way we live our lives! One of the most recent and exciting developments is the revolutionary AI model called ChatGPT. This tool gained a million users just five days after it was made public in November 2022. It has grown a lot since this time, and continues to amaze people all over the world with its seemingly endless use cases! It has proven to be an incredible tool for people all over the world, from students and researchers to entrepreneurs, employers and employees.

Basically, ChatGPT's limitations seem to depend only on the limits of our human creativity. The most incredible part of all is that, at less than a year old, this technology is still in its infancy! The most recent version of ChatGPT, version 4, was released on March 14th, 2023, and has expanded on the capabilities of the last version. Thus, it's not a surprise that people are realizing this technology can be used to make life-changing money, and has already done so for many of its ambitious, early adapters.

You see, what these early adapters understand is that AI technology comes with a very unique opportunity to achieve financial freedom in far less time than it has ever taken in the past. There is good reason to believe that this technology will help countless people make their first million dollars, and you can be among them!

This book will give you the knowledge you'll need to maximize your earnings and make the most out of this powerful technology. We will provide a simple, easy-to-understand guide that will teach you all that you need to know.

The Rise of ChatGPT

The organization responsible for the breakthrough ChatGPT technology is OpenAI. This AI language model is capable of understanding human languages, but what makes it stand out from older chatbots like Google, is its ability to have full-on conversations with users, remember those conversations, and assist users in everything from day-to-day, simple questions to complex problems that take a lot of thought and expertise. But there is more that ChatGPT is capable of including offering individuals, entrepreneurs, creatives, investors invaluable data-driven insights, creative solutions, and essential guidance along their paths to financial success. It really is an all-in-one, amazing tool with unstoppable potential.

How AI can revolutionize the way we make money?

Making a lot of money on a consistent basis is not something that happens by accident. Creating wealth, and most importantly *keeping* wealth and *growing* wealth, are endeavors that demand knowledge, skill, and dedication. It requires setting up systems that will run whether you are sleeping, traveling or even watching TV.

In the past, it would take years of learning, practicing, and trial-and-error to understand and master the art of wealth accumulation. But with the modern advancements in AI technology like ChatGPT, access to sophisticated resources is now available to everyone! Literally, anyone with an internet connection and a phone or laptop can now obtain valuable insights and acquire the same abilities as some of the most renowned financial gurus. Basically, ChatGPT is here to even the financial playing field!

The scope of the book

This book is designed to be your ultimate guide on using ChatGPT to earn your first million dollars! Or, at least create life-changing money. In this book, you will learn many of the ways in which ChatGPT can help you manage your personal finances, create profitable businesses, as well as give you practical tips on how to stay ahead of the AI curve.

With a blend of cutting-edge technology and timeless financial wisdom, this book will equip you with the knowledge and resources you need to become the first millionaire in your family.

In the following chapters, we will walk you through the journey of transforming your financial future by harnessing the power of ChatGPT. From understanding the technology itself to building a successful ChatGPT-based business, this book aims to be your roadmap to wealth. So, get ready. It's time to change your life forever!

Capabilities and limitations

ChatGPT has the ability to engage in human-like conversations, generate creative and coherent text, and adapt to a wide range of topics and tasks, which makes it an invaluable tool for anyone wanting to pursue success. But, even with these remarkable capabilities, there are limitations that we need to be aware of before using ChatGPT to make money.

For instance, the AI model can sometimes misunderstand your input (prompts) or miss subtle nuances when trying to understand context. This may result in responses that are irrelevant or incorrect.

This is why it is so important to review, and double-check the information or task given to you by the chatbot. This helps ensure its accuracy, appropriateness and relevance. With knowledge of both the strengths and weaknesses of ChatGPT, you will be able to make informed decisions on how best to use this powerful technology to make a 7-figure income.

Chapter 2

Principles of Wealth Creation

To effectively use ChatGPT to generate a million dollars, it's important to first understand the underlying principles that determine financial success. In this chapter, we'll explore a few key concepts that form the foundation of wealth-building: the importance of a growth mindset, time management, the role of financial education, the significance of saving, investing, and diversification, and the power of compound interest.

The importance of a growth mindset

A growth mindset is a belief that your abilities, intelligence, and talents can be developed and improved over time. Embracing a growth mindset is crucial for wealth creation, as it encourages you to learn from mistakes and setbacks, embrace challenges and seek opportunities, persist in the face of obstacles, continuously improve and adapt to change.

Time management

Ever wonder why although everyone has the exact same 24 hours in a day, some people are able to achieve so much during each day? You know the type: they make enough money to maintain their impressive lifestyle, they have time for family, for working out, for resting, for traveling and more! Meanwhile, the rest of the population struggles just to establish a proper work-life balance.

Studies have shown that at least 35.2% of all adults in the U.S. report sleeping less than seven hours per night on average. And in 2017, 52% of young professionals were the least likely to factor exercise into their daily lives, stating that they lacked time for it. With so many adults in the U.S. sleeping and exercising properly due largely to feeling overworked, you'd think that at the very least most American adults would at least be financially stable. But that couldn't be further from the truth. As of January 2023, 60% of U.S. adults reported living paycheck to paycheck.

So, why is it that so many people spend so much time working, but can't seem to get ahead? And why are some of the wealthiest people in our society able to earn so much money, while they still have time to live fun, exciting, relaxing and healthy lives?

Well, one crucial aspect of achieving financial freedom and a stress-free life is something called *time management*. Time management is the process of organizing, planning, and effectively using one's time to accomplish specific tasks, goals, and objectives. It involves prioritizing tasks based on their importance and urgency, allocating appropriate time for each task, and using techniques to improve focus and productivity. Good time management enables individuals to work smarter, not harder, allowing people to achieve more in less time. On the other hand, poor time management can lead to burnout and reduced productivity.

Millionaires understand the importance of efficiently managing their time and energy. They plan out their days, weeks, months, and even their entire year well in advance. Generally they will write down what they need to get done each day, and do their best to finish the important tasks before noon. They often divide up their working hours as well as their free time, so they can make sure they are taking adequate vacations, exercising regularly, getting enough sleep, seeing friends and family, etc. They don't waste long periods of time aimlessly scrolling through social media or binging on Netflix —they focus on activities that lead them closer to achieving their goals. In addition, they are comfortable saying no and setting boundaries when necessary. Time management is essential for anyone striving for success but it becomes especially important if you want to become a millionaire.

By managing one's time effectively, one can maintain a healthy work-life balance, reduce stress and ensure sustained effort towards their financial goals. This is something that wealthy, well-adjusted, well-rested and happy people understand. They are always finding ways to save time while maximizing their productivity, instead of simply being busy with time-wasting tasks. Sadly, the majority of people find themselves so overworked because they don't prioritize money-making tasks. They spend too much time on activities that take a long time to complete, and that do not make money. Then, by the end of the day they are burnt out, but they have not made as much money as their rich counterparts.

This is why we need to let go of the idea that working hard always leads to greater wealth. For the most part, it does *not*. If life really worked this way, every hard working person would be rich.

Have you ever sat down to write an email, whether it was on behalf of your company, or it was a message to a client, brand or customer, only to realize you've spent nearly half an hour on it? Have you ever sat down trying to create a product description, write a resume, create a bio for a professional social media profile, etc.? Such tasks can take a long time, and can leave you feeling tired, burnt out and busy, especially if you have a few to complete! But, while these tasks are time-consuming, they are not guaranteed to earn you extra money. And therein lies the problem.

Those with money tend to concentrate their efforts on activities that will generate more wealth- usually the activities that they excel at and are fond of. Employing workers or assistants to take care of duties that take up too much of their time, or which don't directly help bring in extra revenue is often a tactic used by those who are well-off. Fortunately, ChatGPT can provide an answer for anyone who does not have expendable funds to pay an assistant.

Did you know that ChatGPT can be used to:

- Craft emails of all kinds

- Provide customer service to customers and clients

- Craft professional messages to reach out to brands, individuals for potential partnerships and collaborations, etc.

- Write code

- Articles

- Marketing copy

- Stories, poems

- Generate ideas

Take advantage of the time that ChatGPT can save you by treating it as a personal assistant that you don't have to pay. Delegate your arduous tasks to it,so you can focus on the things that will earn you more money while saving you time.

The role of financial education

Financial education is the foundation of any wealth-building journey. It equips you with the knowledge and skills to make informed decisions about your finances. Key aspects of financial education include:

- Understanding personal finance basics, such as budgeting and debt management

- Learning about investment options, including stocks, bonds, and real estate

- Staying informed about economic trends and financial news

- Familiarizing yourself with tax laws and regulations

Saving, investing, and diversification

It's been said that the foundation of wealth creation is the practice of saving, investing, and diversifying your assets, and for good reason. They are foolproof methods that you need to implement into your life no matter how much money you make, in order to gain financial security and see your money grow overtime. Let's take a closer look at these strategies to help you build and protect your wealth.

Saving: The old saying "pay yourself first" is true. It's important to regularly set aside a portion of your income for savings and investments. Even an amount as small as 10% of your pay each week, or each month will compound over time and go a long way in helping you on your journey to financial freedom.

As we are experiencing a global financial downturn, it is wise to save 3 - 6 months worth of income at the least, and set it aside in case of an emergency. Ideally save up to 12 months if possible, but 3 - 6 will allow you some security in case anything should happen in the short-term that hinders your ability to pay your monthly expenses. Ideally, it is wise to open a tax free savings account and put this money there, so that it is growing overtime, and so that you are less likely to spend it spontaneously.

Investing: Once you have your savings in place, you should allocate some of your income into different investments that generate returns over time. There's a common misconception that only rich people invest in stocks, or that you must wait until you are wealthy to buy them. But this is untrue. Even buying a few dollars worth of stocks each week can make a big difference overtime.

When it comes to investing, the simplest option is often the best choice: start with what you know. Invest in the companies you are already familiar with and that you use regularly. For example, if you pay for service from T-Mobile each month, why not invest in their company? The same goes for any bills sent by electricity or hydro providers, or car lenders. Put the money back into something you own a piece of - yourself!

When it comes to stocks, dividend stocks should be considered. With the S&P 500, you have companies that have been steadily increasing their dividends for 25 years or more. Dividend King stocks are another option - these are businesses that have had an increase in dividends for no less than 50 consecutive years. Investing in these types of stocks is a great way to build wealth as not only do you own parts of some of the world's most prosperous companies, but they give regular payouts either monthly or quarterly. As an investor, you experience capital growth while still having a regular and dependable passive income. And honestly, that is one of the major secrets to being a millionaire, creating passive streams of income. The more you have is the better. You may have heard the saying that "the average millionaire has 7 streams of income." And remember what we said about wealthy people valuing their time, and outsourcing work? This is what we meant. They work smarter, not harder. And putting your money to work for you is definitely a smart way to earn money, without tiring yourself out!

If you want to forgo a stock broker, you can take control of your own stock portfolio by downloading any reputable stock trading app, fund your account and begin buying stocks. Many of the stock trading apps also give you the option of purchasing fractional shares (a portion of an equity stock) if you cannot afford, or do not want to own one whole share.

Stop thinking of money as something you have to work to earn. Your money should be working for you. It should be making you money. Investing your money puts it to work, so that it can generate income for you while you sleep.

Diversification: It's important to spread your investments across different asset classes and industries to minimize risk. This could look like investing money in stocks, bonds, silver, and as your wealth increases investing in real estate, gold and other commodities is very wise.

The power of compound interest

Compound interest is the process by which your investments grow exponentially over time, as both the initial principal and the accumulated interest earn interest. The power of compound interest lies in its ability to:

- Accelerate wealth accumulation

- Reward long-term investing and patience

- Mitigate the impact of market fluctuations

- Enable you to reach your financial goals more quickly

Understanding and applying these principles of wealth creation will prepare you to effectively leverage ChatGPT in your journey towards financial success. In the following chapters, we'll explore how to use ChatGPT as a tool for personal finance management, business growth, and investment analysis, and more in order to lead you towards the path of becoming a millionaire, all while keeping these foundational concepts in mind.

How ChatGPT can be used to save you time:

1. Ask for time management tips: Request specific advice or strategies for managing your time more effectively. It can provide suggestions based on proven techniques and best practices.

2. Goal setting and prioritization: Share your goals, and it can help you break them down into smaller, manageable tasks. It can also help you prioritize these tasks based on their importance and deadlines.

3. Task scheduling: Ask it to help you create a daily, weekly, or monthly schedule that allocates time for your tasks, work, and personal life. This can help you stay organized and maintain a healthy work-life balance.

4. Productivity techniques: It can introduce you to various productivity methods, such as the Pomodoro Technique, time blocking, or the Eisenhower Matrix, and help you determine which techniques might work best for your needs and

preferences.

5. Identifying distractions and time wasters: Discuss your daily routine, and it can help identify potential distractions or time-wasting activities that you can minimize or eliminate to increase your productivity.

6. Task delegation: it can provide guidance on how to delegate tasks effectively, freeing up your time to focus on high-priority activities.

7. Resource recommendations: it can recommend books, articles, podcasts, or courses related to time management and productivity. These resources can provide additional insights and strategies for managing your time more effectively.

8. Reminders and motivation: it can provide reminders and motivational support to help you stay on track with your time management goals.

Remember that, as an AI, it cannot directly interact with your devices, calendars, or applications. You'll need to implement any recommendations or changes manually. But ChatGPT is an excellent assistant that can guide and support you in improving your time management skills, so feel free to ask any questions or request additional information.

5 Prompt examples you can ask ChatGPT in order to manage your time better:

1. "What are some effective strategies for prioritizing tasks and managing my workload?"

2. "Can you explain the Pomodoro Technique and how it can help me stay focused and productive?"

3. "How can I create a daily routine that balances work, personal life, and self-care?"

4. "What are some tips for overcoming procrastination and staying motivated to complete tasks?"

5. "How can I set and achieve SMART goals to improve my time management skills?"

5 Prompt examples you can ask ChatGPT in order to begin successfully investing in the stock market:

1. What are some important factors I should consider before investing in the stock market, and how can I determine which stocks are a good in-

vestment?

2. Can you recommend any specific strategies or tools for analyzing market trends and identifying profitable stocks to invest in?

3. What are some common mistakes that beginners make when investing in the stock market, and how can I avoid them?

4. How do I develop a diversified investment portfolio that balances risk and potential returns across different sectors and asset classes?

5. Are there any specific industries or sectors that you would recommend for long-term investment growth? Why?

Chapter 3

ChatGPT for Personal Finance Management and Finance Coaching

The power of ChatGPT as a financial coach and finance manager cannot be underestimated. It is an AI-powered program that can provide advice, guidance and even insights into your finances. This technology can act like a personal assistant for money management, helping to keep track of expenses, investments, budgeting and more. ChatGPT can also help you identify areas where you could potentially save money, analyze your income and expense trends and offer up actionable steps to improve your financial situation. The technology is designed to constantly learn from the user's data, so it can come up with ideas on how to better manage their finances in the long term. Overall, ChatGPT is quickly becoming a powerful tool for anyone looking to take control of their finances. In this chapter, we'll be taking a look at how ChatGPT can act as a powerful tool to help you not only manage your money better, but how it can offer you finance coaching. Both of these are essential to your journey of becoming a millionaire.

Budgeting with ChatGPT

Creating and maintaining a budget is essential for managing your finances effectively. ChatGPT can help with:

Setting up a budget template: ChatGPT can generate a personalized budget template based on your income, expenses, and financial goals.

Categorizing expenses: ChatGPT can help you classify your expenses into categories, making it easier to track and manage your spending.

Identifying areas for improvement: By analyzing your budget, ChatGPT can suggest areas where you could cut back on expenses or increase savings.

Budget adjustments and support: ChatGPT can help you make adjustments to your budget as your financial situation or goals evolve. It can serve as an accountability partner, offering regular reminders and updates on your budget progress.

Here are some effective prompts you can use for budgeting:

1. "ChatGPT, could you help me create a monthly budget based on my income of $X, fixed expenses of $Y, and financial goals such as saving 20% of my income, paying off debt, and allocating 10% for entertainment?"

2. "ChatGPT, I'd like to review my current budget and identify areas for improvement. Here's a breakdown of my monthly income and expenses: [provide details]. Can you analyze my spending and suggest ways to optimize my budget to achieve my financial goals?"

3. "ChatGPT, I'm new to budgeting and need some guidance on setting up a budget that works for me. My monthly income is $X, and my main expenses are housing, utilities, groceries, transportation, and entertainment. Can you provide a

step-by-step guide to creating a budget that helps me save money and reach my financial goals?"

Tracking expenses and identifying trends

Keeping track of your expenses is crucial for maintaining a healthy financial life. ChatGPT can assist with:

Expense tracking: ChatGPT can help you record and track your daily expenses, providing you with an up-to-date overview of your spending.

Analyzing spending patterns: By reviewing your expense data, ChatGPT can identify trends and patterns in your spending habits.

Uncovering hidden expenses: ChatGPT can help you discover overlooked expenses that could be draining your finances.

Debt Management

According to Credit Karma, individuals making $184,000 per year or more, have the most credit card debt at $12,600 on average. The total credit debt Americans held in 2021 was $784.5 billion, and with the inflation that has happened over the last couple years, you can only imagine what the collective debt is today. No wonder so many people are caught up on the hamster wheel, chasing extra income while running away from debt, and getting nowhere thanks to inflation.

Here's the honest truth, if you want to become a millionaire in today's financial landscape it is possible. We live in such an amazing time full of opportunities and potential for new businesses, for easier exposure for yourself and your brand, and for better partnership opportunities! New millionaires are being made every year!

But, in order to become one of the next millionaires, you need to get rid of your credit card debt. The interest rates are increasing constantly and this means that you are being forced to pay a higher and higher percentage on the money you are borrowing via credit cards. Pay them off and then never spend more money on a credit than you are capable of paying back by the end of the month.

Of course, if you are currently drowning in credit card debt, you need to put all your focus into earning more money so that you can pay off your debt, while spending as little of your remaining credit balance as possible. Ideally, in order to have good credit you want to carry a balance that is less than 30% of your credit card's limit.

So, for example, if your credit card limit is $1000, you don't want to carry a balance that is over $300. When you follow this guideline consistently, credit card companies will see that you have money, and that you don't need their money. This will incentivize them to offer you more money in the form of an increased credit limit. On the other hand, running up your credit cards actually makes banks nervous about having you as a customer. It shows them that you are not good with money, and that you may not be capable of paying them back their money. Funny how that works, right?

In short, spend less than 30% of your credit card's limit, and never miss payments. As well, try to never spend more than you can pay back by the end of the month. Follow these rules and you will be well on your way to being debt free, and having good credit!

Good credit will open so many doors for you. For instance, it can be a major advantage when it comes to buying big-ticket items like a house or car. With good credit, you can benefit from lower interest rates on mortgages and auto loans, better loan terms, easier loan approval, and increased access to lenders and loan options. All of this could save you thousands of dollars in the long run, so it's worth taking the time to pay off your debt and build your credit score.

Here's how ChatGPT can offer guidance on managing and reducing your debt effectively:

    1. Debt analysis: ChatGPT can help you assess your

current debt situation, including interest rates, monthly payments, and remaining balances.

2. Repayment strategies: ChatGPT can suggest various debt repayment strategies, such as the snowball or avalanche methods, to help you pay off your debts more efficiently.

3. Debt consolidation: ChatGPT can provide information on debt consolidation options, such as balance transfer credit cards or personal loans, to potentially lower your interest rates.

4. Credit score improvement: ChatGPT can offer tips on improving your credit score, which can result in better loan terms and lower interest rates in the future.

Saving and Investing

ChatGPT can help you build a robust savings and investment plan:

1. Emergency fund: ChatGPT can guide you in establishing an emergency fund, determining the ideal amount to save and suggesting strategies for building it up.

2. Savings goals: ChatGPT can help you set short- and long-term savings goals, such as saving for a vacation, home purchase, or college education.

3. Investment strategies: ChatGPT can offer insights into various investment strategies, such as passive investing, dividend investing, or value investing, to help you grow your wealth.

4. Risk management: ChatGPT can provide guidance on managing investment risks, including diversification, asset allocation, and risk tolerance.

Generating personalized financial advice

ChatGPT can offer valuable financial advice tailored to your unique circumstances and goals:

Answering financial questions: ChatGPT can provide information and insights on a wide range of personal finance topics, such as debt management, retirement planning, and tax strategies.

Goal setting: ChatGPT can help you establish realistic financial goals and develop a plan to achieve them, such as buying a home, starting a business, or achieving financial independence.

Developing a financial plan: ChatGPT can generate a comprehensive financial plan that outlines your saving, investing, and wealth-building strategies. It can also assist you in creating a detailed action plan to achieve your financial goals, breaking them down into manageable steps.

Progress tracking: ChatGPT can help you monitor your progress toward your financial goals, providing encouragement and suggesting adjustments as needed.

Continued support: ChatGPT can be used as a resource for continuous financial guidance and support, which helps you stick to your financial plan easier and adapt to unpredictable circumstances that may arise.

Celebrating milestones: ChatGPT can celebrate your financial milestones with you, reinforcing your motivation and acknowledging your hard work and dedication.

Analyzing investment opportunities

ChatGPT can be a helpful tool for researching and evaluating different investment strategies and opportunities. It can help you collect data on various investment opportunities, compare different investments based on risk factors, potential returns, and fees, evaluate the risks and rewards of certain investments, and track your investment performance over time. In short, ChatGPT is an invaluable financial ally that can assist you in making the best possible investments decisions.

By leveraging ChatGPT in these areas of personal finance management, you can gain valuable insights and guidance to help you make smarter financial decisions and move closer to your wealth-building goals.

Let's consider a real-world example of how ChatGPT can assist you in managing your personal finances:

Imagine you're a young professional looking to gain control over your finances and work towards a secure financial future. ChatGPT can help you create a personalized budget, tracking your income and expenses while suggesting ways to cut costs and develop better financial habits.

As you work to pay off student loans or credit card debt, ChatGPT can offer guidance on repayment strategies, debt consolidation options, and improving your credit score. To build your savings and investments, ChatGPT can assist you in setting goals, choosing investment strategies, and managing risk.

When it comes to retirement planning, ChatGPT can help you set realistic goals, maximize your retirement savings, and create a sustainable withdrawal strategy. Finally, ChatGPT can support you in setting and achieving your financial goals, from purchasing a home to starting a business or reaching financial independence.

By utilizing ChatGPT as your personal finance coach, you'll gain valuable insights, guidance, and encouragement to help you navigate the complexities of personal finance and achieve long-term financial success. With the power of AI technology at your side, you'll be well-equipped to make informed decisions and take control of your financial future.

Retirement Planning

It's true, retirement planning is not a sexy topic, which is why people tend to avoid it. The rather sad reality is that we will not be young forever. We will not have the ability, strength or energy to work as much or as hard as we do now, thirty, forty, fifty years from now. In order to become a millionaire who lives a long and fulfilling life, a life where you are not living on a small and fixed income in your golden years, you should be thinking about retirement planning right now!

Retirement doesn't have to be limited to those aged 65 and over, nor does it have to be the end of a person's life or income. Many people in their 30s, 40s and 50s are learning how to save money successfully and generate money passively through assets (such as dividend stocks, digital books, property etc.) so they can achieve an early retirement and live more life! After all, who wants to wait until they are 65+ to travel the world regularly, or spend each day however they want? When you look at retirement like this, the point in your life when you choose to stop working, you realize this is something we should all be aiming for: the ability to spend your days however you want, true time freedom so you can live more life!

So, no matter when you want to retire - early in life, or later in life - you can use ChatGPT as a valuable resource in planning for a comfortable retirement. Here are a few ways that ChatGPT can help you with retirement planning:

1. Retirement goals: ChatGPT can help you set realistic retirement goals, such as your desired retirement age, income, and lifestyle.

2. Retirement savings: ChatGPT can suggest strategies for building your retirement savings, including maximizing employer-sponsored plans and utilizing individual retirement accounts (IRAs).

3. Social Security benefits: ChatGPT can provide information on Social Security benefits and help you determine the optimal claiming strategy.

4. Retirement withdrawal strategies: ChatGPT can

offer guidance on creating a sustainable with-drawal strategy during retirement, ensuring your nest egg lasts throughout the rest of your life.

5. Passive income opportunities: ChatGPT can provide you with plenty of ways you can create passive income that you can live off of for years into the future.

Chapter 4

ChatGPT for Investing Success

I n this chapter, we'll explore how ChatGPT can be used to help you make informed investment decisions by providing valuable insights, data analysis, and research support. We'll look at a few major areas where ChatGPT can help improve your investment success, which will bring you closer to becoming a millionaire and living a life of financial freedom.

Understanding Market Trends

You can use ChatGPT to gain valuable insight into market trends and economic indicators, which helps you make better-informed investment decisions over time.

Economic analysis: ChatGPT can help you understand the larger economic landscape. It can do this if you, as the user, as it to summarize "key indicators" like GDP growth, inflation, and unemployment rates.

Sector trends: ChatGPT can give you insight into not only the performance, but the outlook of different industry sectors. This allows you to notice opportunities and any potential risks. So that you are always making the best decisions.

Market sentiment: ChatGPT can gauge market sentiment by analyzing news, social media, and other sources, helping you anticipate potential market movements.

Global perspective: ChatGPT can provide you with updates on international markets and economies. This helps you diversify your investments across different regions of the world.

Individual Stocks Research

You can use ChatGPT to act as your personal research assistant when evaluating individual stocks to buy. In fact, prompts such as *"act as if you are an Investment Researcher,"* can be effective in training the AI model to provide the most helpful information concerning your stock questions.

Company analysis: ChatGPT can help you analyze a company's financial health, competitive positioning, and growth prospects.

Financial statement review: ChatGPT can assist you in reviewing financial statements, highlighting key metrics such as revenue growth, profit margins, and debt levels.

Valuation assessment: ChatGPT can help you determine a stock's valuation using methods such as price-to-earnings ratio, discounted cash flow analysis, or dividend discount models.

Risk analysis: ChatGPT can identify potential risks associated with a stock, such as market volatility, industry-specific factors, or management concerns.

Assessing Real Estate Opportunities

Real estate is one of the best wealth building assets you can own in order to generate more passive income in your life, and create wealth. In fact, the words of Andrew Carnegie, one of the wealthiest people in history, still holds true today, that "90% of all millionaires become so through owning real estate." And while entrepreneurship also contributes largely to creating self-made millionaires these days, real estate will always be a safe, and easy bet as well.

Most American millionaires invest in real estate, and it's completely understandable. This investment strategy has a range of benefits that make it an attractive option for wealth building in your own life. Here are some of the main ways that real estate creates millionaires:

Stable cash flow: Real estate investments, like rental properties, provides you with a steady stream of income, which ensures financial security. It also allows investors to plan their future expenditures better.

Tax advantages: Real estate investments offer a variety of tax benefits, including depreciation, mortgage interest deductions, and capital gains tax deferral through 1031 exchanges. These advantages can help investors maximize their returns and minimize tax liabilities.

Appreciation: Over time, property values tend to increase, at a rate that is even faster and higher than even most stocks! This allows investors to build wealth through capital appreciation. This appreciation can also serve as a hedge against inflation, and protect an investor's purchasing power. This appreciation can also be taken out of your property, and used to purchase other things, like down payments on other rental properties.

You can see how this can help you purchase more and more real estate over time, and earn huge amounts of passive income, eventually leading to quite a bit of wealth.

The kind that can even be passed down throughout your family, called generational wealth. Literally, owning real estate can end the generational curse of poverty in your family!

Leverage: Real estate investments offer the opportunity to use leverage, meaning investors can acquire properties by borrowing funds and using a smaller amount of their own capital. This allows them to expand their portfolios and increase their potential returns.

Diversification: Including real estate in an investment portfolio helps diversify risk and reduce overall volatility. By spreading investments across various asset classes, investors can minimize the impact of a single underperforming asset on their overall returns.

Control: Real estate investments give investors more control over their assets compared to other investment options, such as stocks or bonds. They can actively manage their properties, make improvements, and influence the value of their investments.

Tangible asset: Real estate is a physical asset that can be seen, touched, and improved upon. This tangible nature provides a sense of security and satisfaction that many investors find appealing.

So how does ChatGPT play into all of this? ChatGPT can provide support in evaluating real estate investments among other things.

Market research: ChatGPT can help you research local real estate markets, identifying trends in property values, rental rates, and inventory levels.

Property analysis: ChatGPT can assist you in analyzing individual properties, considering factors such as location, condition, and potential return on investment.

Financing options: ChatGPT can provide information on various financing options for real estate investments, including mortgages, home equity loans, or private lending.

Cash flow projections: ChatGPT can help you create cash flow projections for rental properties, estimating potential income, expenses, and profitability.

General Advice: ChatGPT can act as an advisor in general, answering your standard questions about the real estate industry. What to do, what not to do, things to avoid, common mistakes, etc.

Evaluating Alternative Investments

ChatGPT can offer insights into alternative investments, such as cryptocurrencies, commodities, and even art.

Asset overview: ChatGPT can provide an introduction to various alternative assets, explaining their unique characteristics and potential benefits and risks.

Market analysis: ChatGPT can assist you in analyzing the markets for alternative investments, identifying trends, opportunities, and challenges.

Investment strategies: ChatGPT can suggest investment strategies tailored to specific alternative assets, such as dollar-cost averaging for cryptocurrencies or diversifying within a commodity portfolio.

Risk management: ChatGPT can offer guidance on managing the risks associated with alternative investments, such as implementing stop-loss orders or maintaining a well-diversified portfolio.

Monitoring Your Portfolio

ChatGPT can support you in keeping track of your investment portfolio and making adjustments as needed.

Performance tracking: ChatGPT can help you analyze your investments' performance while also providing you with regular updates.

Portfolio analysis: ChatGPT can access your portfolio's composition, risk level and exposure, and diversification. It can then offer tips on how to make it better.

Rebalancing recommendations: ChatGPT can help you decide when and how to rebalance your portfolio for the optimal return on investment that fits with your desired risk level.

Here's an example of how ChatGPT can help you on your investment journey.

Let's say you're a new investor looking to build a diversified portfolio. ChatGPT can provide you with a better understanding of current market developments, and obtain insight into different industries. Based on your risk tolerance and investment goals, ChatGPT can suggest a mix of stocks, bonds, real estate, and alternative investments.

Let's consider an example of how ChatGPT can help you on your investment journey. Imagine that you are a brand new, beginner investor seeking to build a well-diversified portfolio. With ChatGPT, not only will you get clear information about the current market trends, but also in-depth perspectives on each industry. Based on your financial goals and risk profile, ChatGPT can provide you with tailored recommendations for different assets such as stocks, bonds, real estate, and alternative investments.

As you evaluate individual stocks, ChatGPT can provide company analysis, financial statement reviews, valuation assessments, and risk analysis to help you make informed decisions. If you're interested in real estate, ChatGPT can assist you in researching local markets, analyzing properties, and exploring financing options.

For those considering alternative investments, ChatGPT can offer an overview of various assets, market analysis, and investment strategies tailored to each type of asset. Finally, ChatGPT can help you monitor your portfolio, track its performance, and make adjustments as needed to ensure you remain on track to achieve your financial goals.

By leveraging ChatGPT as your investment analysis partner, you'll be well-equipped to navigate the complex world of investing and make more informed decisions that align with your financial objectives. The power of AI technology, combined with your own intuition and judgment, can help you build a successful investment portfolio and achieve long-term financial success.

Chapter 5

ChatGPT for Building a Successful Online Business

D id you know that half of all millionaires are either business owners or are self-employed? It's true, instead of working a job you hate, begging for a raise and saving every penny in hopes of becoming a millionaire, you can take your future into your own hands and create your own financial freedom! In fact, your odds of becoming a millionaire increase exponentially when you go this route.

This chapter is where things get fun! It's time to use the potential of ChatGPT to create a profitable business.

Many new millionaires have created their wealth through entrepreneurship via an online business. An online business is a great way to generate wealth because you can choose a business model with little to no barrier to entry – meaning it is simple to start and does not require special training. As well, when your business is online it is always open, even when you are asleep, due to the constant availability of the internet. And finally, it is highly scalable, meaning that many online businesses have huge growth potential considering you can always use virtual assistants, apps and programs to take on time-consuming or difficult tasks in your business, so that you can focus on your business' profits, growth and expansion.

In this chapter, we'll cover some core components of building a successful business using ChatGPT. You'll see how you can use ChatGPT to find out how to best meet the market's needs and wants, create the best products and services people want to buy, and understand the different ways you can monetize your business, all while abiding by legal and ethical considerations.

Identifying Market Needs and Opportunities

ChatGPT can be a powerful tool in understanding and responding to market demands.

Market research: With ChatGPT, you can uncover the unexplored needs of your potential customers. You can create prompts asking it to help you spot trends before they become popular, so you can earn the most money. And you can use it to identify any weak points in the services or products currently on the market, so that you can make your products or services even better than theirs.

Competitor analysis: ChatGPT can help you analyze your competitors, and understand their strengths, weaknesses, while providing you with opportunities for differentiation.

Customer insights: ChatGPT can also help you gather feedback from customers directly, so that you can make informed decisions to tailor your products/services to their specific needs, preferences and struggles. Through Chat-GPT, your business will gain an edge in navigating the ever-changing marketscape.

Opportunity evaluation: ChatGPT can help you assess the viability of identified opportunities by considering factors such as market size, growth potential, and competitive landscape.

Designing and Launching ChatGPT-powered Products and Services

Why not leverage ChatGPT's capabilities to create innovative products and services?

Product ideation: ChatGPT can help you brainstorm and refine ideas for ChatGPT-powered products or services, ensuring they address real and current market needs and offer value to customers.

Prototype development: ChatGPT can assist you in creating prototypes or mockups of your product or service, providing feedback and suggestions for improvement throughout the process.

Testing and validation: ChatGPT can help you design and execute testing plans to validate your product or service, gathering user feedback and iterating on your design based on their input.

Launch strategy: ChatGPT can offer guidance on crafting an effective launch strategy, including marketing, sales, and customer support plans to ensure a successful market entry.

Monetization Strategies and Revenue Streams

Develop revenue-generating strategies for your ChatGPT-based business:

Pricing models: ChatGPT can help you explore different pricing models and strategies, such as subscription-based, freemium, or pay-per-use, to determine the most suitable approach for your business.

Sales channels: ChatGPT can assist you in identifying and optimizing the most effective sales channels for your products or services, such as online marketplaces, direct sales, or partnerships.

Marketing and promotion: ChatGPT can offer guidance on creating and executing marketing and promotional campaigns to attract customers and generate sales.

Ancillary revenue streams: ChatGPT can help you identify potential ancillary revenue streams, such as licensing, affiliate marketing, or sponsored content, to further monetize your business.

Navigating Legal and Ethical Considerations

It's important that your ChatGPT-based business operates within legal and ethical boundaries, and ChatGPT can help you to do this.

Intellectual property: ChatGPT can provide you with information concerning intellectual property rights, like patents, trademarks, and copyrights, to help you protect your business while also avoiding copyright infringement.

This can look like checking to make sure your business name is not already owned by another company. It can also look like making sure the products (physical or digital) you create, or sell, are not infringing on the trademarks of another company.

For instance, a few years ago people realized you can make a lot of money selling books on Amazon. Even the average person who is not a prolific writer could make money selling books, by focusing on "low content books" - notebooks, lined journals, prompt journals, sketchbooks, coloring books, etc. Needless to say, thousands of people flocked to the business model, and many found success, earning life-changing income.

Now, because low-content books are so easy to create, these self-publishers constantly had to hop on new trends, and publish countless amounts of books, because they were saturating niches and keywords. They soon realized that there was a huge demand for a journal known as a "bullet journal." So, because it was such a highly-searched term, many "low-content" self-publishers began publishing lined journals on Amazon, and placing "Bullet Journal" in their book titles.

Over time, all of their accounts ended up being flagged and their books were removed from the platform. Why? Because they infringed on the copyright of Ryder Carroll, the original owner of the *Bullet Journal Method*. That's right, they were all accidentally using a copy-righted term in their book titles, that was flagging their accounts and compromising their entire self-publishing careers!

With ChatGPT, you can create a prompt asking the AI model to check for copyrights prior to publishing a book, or launching a business of any kind. This way, you can avoid unfortunate situations like the one mentioned above, in your business.

However, for the latest, up-to-date information, it is also important to double-check the responses ChatGPT gives you with more up-to-date websites like:

U.S. Copyright Office: https://www.copyright.gov/

Copyright Clearance Center: https://www.copyright.com/

Google Books: https://books.google.com/

Data privacy and security: ChatGPT can offer guidance on complying with data privacy and security regulations, such as GDPR or CCPA, to ensure your business handles user data responsibly.

Ethical AI practices: ChatGPT can help you navigate the ethical implications of AI technology, promoting transparency, fairness, and responsible use in your business.

Industry-specific regulations: ChatGPT can assist you in understanding and adhering to any industry-specific regulations or guidelines that may apply to your business.

Let's take a closer look at a real-world example of how ChatGPT can be put into action to build a profitable online business.

Firstly, one of the easiest and most profitable businesses you can start in this day and age is an online business. It's particularly easy to start on a marketplace like Etsy, Amazon, eBay, Teespring, Redbubble, Zazzle, etc. due to the "warm traffic" - ready-to-buy customers who frequent them.

You can sign up for free on any of these websites, but personally I love Etsy for beginners looking to make money online using the print-on-demand business model. It is what I used and made regular $10k months in under 2 years of working on my business part-time. It's a low-barrier to entry business model, meaning there are very little upfront fees, and you can sign up for free right away.

So, let's Imagine you'd like to start a profitable Print-On-Demand Etsy shop business. And yes, print-on-demand is still a great way to make life-changing money. In fact, many of the top-earning Etsy stores that make +$1 million annually are print-on-demand stores.

Let's imagine that you've already signed up to Etsy. The next thing you could do to build a successful shop is use ChatGPT to:

- identify profitable niche markets that you could target and create products for

- generate a list of profitable (low competition, highly searched) print-on-demand product ideas that people in your niche would enjoy

Once this is complete, you will need a print-on-demand partner, a business where you can create your unique designs on white label products. They will ship out each order you receive from Etsy customers, on your behalf. That is what makes print-on-demand such a superior business model for selling physical products in this day and age. You don't have to hold on to any inventory, and you can jump on trends much faster.

Here are some print-on-demand partners that automatically interrogate with Etsy, including Printify, Printful, Printed Mint, Gooten, Shirtly, Print Aura, Art of Where, and TeeLaunch. But in reality, the sky's the limit. There are countless P.O.D companies you can use to fulfill Etsy orders, just be aware that you will have to fulfill those orders manually if they are not official Etsy partners.

If you'd prefer a digital product business instead, where you are selling digital downloads, also known as printables, you can use ChatGPT to research popular digital download ideas and low competition niches. You can further validate any responses the AI model gives you with Etsy research tools like Everbee, which offers more uptodate information on bestsellers, their monthly sales, etc.

A digital product based business works best for people who enjoy creating art, templates of all kinds, and even ebooks! Honestly, both physical and digital products can earn you a lot of extra income, enough to replace your day job if that is what you desire. But it is slightly easier with physical products in my experience, because you can charge more per item depending on what you are selling. Regardless, why not try both and see which one works for you? Then, double down on the products that are selling!

Once you've decided on your business model (physical products, or digital products) you can refine your product ideas and ChatGPT could help you craft a compelling business name, Etsy about section, and product descriptions.

Creating Your Product Designs

The print-on-demand business model - both physical and digital products - is ideal for people who are creative, and enjoy the creation process. That being said, you don't have to be an amazingly skilled graphic artist to make profitable product designs. And don't assume you have to create each design all alone either! We now live in a world where AI has given everyone access to artistic skills at the click of a button, when you use the right prompts. Of course, you can use incredible online platforms like Creative Fabrica, and Creative Market, Canva, Design Bundles, TheHungryJpeg, Vexels, etc. where you can buy fonts, images, workbook pages and more for commercial use on products.

I like to use Creative Fabrica, due to its vast array of fonts and graphics, and I create new product designs in the online program called Canva.

If you are a little more tech savvy, or at least have the ambition to learn, then do not rule out the types of artistic designs that other AI models can create for your print-on-demand business as well. For instance, you can learn to use the popular text-to-image AI model called *Midjourney*. Midjourney helps people, of any artistic level, create stunning and unique designs.

In order to craft effective text-to-image prompts for Midjourney, simply learn from Youtubers who provide prompt ideas, scroll through the Midjourney Reddit groups and take notes on the prompts they recommend, or scroll through Midjourney itself and check out the artwork of fellow prompt engineers and take notes on their prompts. Do this until you feel comfortable combining these prompts and using them in your own AI artwork. Eventually, you will even begin coming up with your unique prompts, through trial and error, and you will have gained yet another skill in the world of AI technology.

It really does not take a long time to gain these skills either. Just sign up under the free plan, and get practicing. Take your best designs and test them on P.O.D. products. I know plenty of people who are making a lot of money right now using this method and selling apparel.

SEO

SEO (Search Engine Optimization) is essential when creating an online business, whether on a marketplace like Etsy, Amazon, eBay, etc. or your own platform, for several reasons:

Visibility and discoverability: Good SEO practices help improve your business' ranking in search engine results. For instance, on Etsy, proper SEO is important in order to rank higher in the search engine. The higher your store ranks in the algorithm, the more likely a potential customer is to find your store when they search for an item. For instance, if your store sells cute butterfly-themed home decor, and a potential buyer types "butterfly decor" into the Etsy search engine, so long as you practice good SEO, your butterfly-themed home decor will show up as some of the first search results that potential customer sees. This gives you a higher chance of selling your items.

A higher ranking on Etsy, or any website where you sell your products, will improve your business' credibility and trust factor. Since online businesses that rank higher in search engine results are often perceived as more credible and trustworthy by users. A higher ranking signals that your website is relevant and authoritative, which can influence a user's decision to engage with your business.

ChatGPT can conduct keyword research for your products, create attractive product titles, and product descriptions, which will in turn, optimize your Etsy listings and increase your sales. This is true for any website, including if you start or currently own a Shopify store. You can ask ChatGPT to generate meta descriptions, among many other things as well.

If using ChatGPT on your own website, like Shopify, using it as a tool to optimize your SEO will help drive organic traffic (unpaid visitors) to your website. Organic traffic is valuable because it tends to have higher engagement and conversion rates compared to paid traffic, as users are actively searching for information or products related to your business.

Keep in mind that marketplaces also act as search engines. They are simply search engines for buying things.

Thus, search engines are continuously being updated to provide better search results, and Etsy is no different. That is why staying up-to-date with SEO best practices, through the help of tools like ChatGPT, ensures that your online business remains visible and relevant to users, even as search engines evolve. This is how you build a long-lasting business!

Imagine you now decided to take this print-on-demand Etsy business, off the platform as well, and create a Shopify store for it, in order to maximize your profits. You could ask ChatGPT to help you come up with a business plan, complete market analysis, marketing strategy and financial projections. You could ask it to come up with a list of your successful competitors as well, so you can study their stores and see what works for them. Afterall, success leaves clues, and the best way to become successful is to emulate (not copy) what already works. When it comes to marketing your brand, ChatGPT can create engaging content for your Instagram, Tiktok, Facebook Page, blog and any other social media. It can brainstorm creative marketing campaigns, and assist with keyword research for SEO and PPC advertising. To streamline your operations, ChatGPT can help you develop SOPs, prioritize tasks, and suggest opportunities for automation.

Finally, ChatGPT can enhance your customer service by creating an informative FAQ section, generating personalized responses to customer inquiries, and even being integrated into a chatbot for instant support.

In this way, ChatGPT can be an invaluable ally for entrepreneurs in their journey to build and grow a successful business. By harnessing the power of this cutting-edge AI technology, you'll be well-equipped to navigate the challenges and opportunities that come with entrepreneurship.

By following these steps and leveraging ChatGPT's capabilities, you can build a successful ChatGPT-based business that addresses market needs, offers innovative solutions, and generates substantial revenue. As you navigate the complexities of entrepreneurship, ChatGPT can serve as a valuable partner, providing insights, guidance, and support throughout the process.

As your business grows, you'll want to continue refining your strategies and exploring new opportunities to stay ahead of the competition. Remember to stay engaged with your customers, as their feedback can provide invaluable insights into areas where your products or services can be improved or expanded upon. It's so important to remain informed about the latest AI tools, and the ChatGPT platform. This is how you become an early adopter of the latest opportunities. And studies have proven that early adopters are often wealthier and have more authority in their chosen field, than the rest of the public. That's why, by staying on top of the latest developments, your business will stay at the forefront of innovation, and consistently adapt to meet the ever-growing needs of the market.

As you can see, creating a successful business using ChatGPT can be a powerful way to become a millionaire. Following the steps listed above can have you well on your way to achieving financial success. With ChatGPT as your trusted partner, you'll have access to a wealth of knowledge and resources at the tips of your fingers. This AI tool literally cuts the time it would normally take to achieve your financial dreams in half!

Prompts you can use to help you build a successful business on Etsy

1. Can you provide a step-by-step guide on how to set up a successful Print-On-Demand business on Etsy?

2. What are the key strategies to build a strong brand for a P.O.D Etsy business?

3. How can I optimize my Etsy listings for better visibility and attract more customers to my

chat-based business?

4. What customer service best practices should I follow to ensure the success of my chat-based Etsy business?

5. How can I effectively use social media marketing to promote my Etsy store and increase sales?

6. Can you help me write impactful product titles and descriptions for my products to stand out from the competition?

7. What are the most common challenges faced by P.O.D. Etsy businesses and how can I overcome them?

8. How can I leverage Etsy's advertising and promotional tools to boost my Print-On-Demand business's visibility and sales?

9. What are some essential tips for managing inventory and order fulfillment for a successful P.O.D -based Etsy business?

10. Can you provide advice on setting competitive pricing and creating enticing promotions for my

P.O.D Etsy business to maximize profits?

11. Bonus: It would be quite crazy of me not to mention *#HustleGPT* and its creator Jackson Greathouse Fall on Twitter. He created his own custom prompt on ChatGPT where he told the AI model it is now called #HustleGPT. He told it that he is its human partner who will follow all its directions. He then tasked the AI model with turning $100 into as much money as possible, in as little time as possible, while not breaking the law. So far his account has blown up overnight and he has amassed a huge follower base of ambitious entrepreneurs, new and old, who are aiming to make more money using the same prompt. Check him out, engage with his community. Give this method a try, and see what ChatGPT can do for you.

Bonus Business Idea: eBooks

As I previously stated in this book, I have personally had great success with both Etsy Print on Demand, as well as self-published books on the platform Amazon KDP, when I first began trying to make money online back in 2021.

In fact, I was able to earn an inspiring amount of monthly income, all profit, just three months after starting my self-publishing business venture. At that time, I had less than 30 books to my name. These books fell under the categories of both low-content (journals, coloring books etc.) and high content (novellas, novels, etc.). I should add that I only worked on this business part time.

That is why I know anyone can replicate my results with some dedication, time, proper niche and keyword research, and commitment. The more you put into any business, the more you will get out. So work hard, or hire a ghostwriter, and craft high quality books. Use ChatGPT to help you decide on the best niches, and keywords. Then, once your book is published, it will act as a wealth-generating asset for you, bringing in money for you every month. Do not spend that money on bills, etc. At least not yet. You do not want to drown your business, before it can even get off the ground.

Instead, you want to reinvest the profits back into your self-publishing business. Use profits to run ads, or purchase high-quality book formatting software like Atticus or Vellum, etc. Use your profits to invest high-quality book covers, so your books stand out amongst others in your niche. Be wise with your money, and make sure that you are putting it to work for you. That is how you create a business that is constantly generating passive income for you, even while you sleep!

Starting Self-Publishing with Amazon KDP and Selling Books

Here is a step-by-step guide on how to use ChatGPT to start a self-publishing business with Amazon KDP selling books:

Research profitable niches and keywords: To begin, ask ChatGPT to identify profitable niches, keywords and topics that are in demand and have potential for profitability. Further validate its responses using free tools like Amazon Best Sellers and Google Trends, or paid tools like KindleSpy, Publisher Rocket, or K-lytics to analyze popular categories, genres, and keywords. Look for niches with a strong reader base and less competition.

Create high-quality content: Ask ChatGPT to help you write well-structured, engaging, and informative book chapters that provide value to your target audience. You can use ChatGPT to ensure that your content is well-researched, and well-written.

Book outlines and Ideas: You can ask the AI model to create a book outline for you, and ask it to brainstorm chapter ideas, and anything else you may need. You can ask it to "act as if you are a bestselling author," and it will produce higher quality writing content you can use for your book. You can use ChatGPT to make sure that your work is free from grammatical errors. Or, you can hire a professional editor to help with proofreading and editing.

Please be mindful that ChatGPT is just a tool, and should not replace you (or a professional ghostwriter) but it can help brainstorm, and flesh out your ideas. It is like having a little writing assistant at your side.

Check for trademarks and copyrights: Ask ChatGPT to generate a list of appealing book titles you can choose from. Now, before proceeding, you can use ChatGPT to help make sure your book's title, content, and images do not infringe on any copyrights or trademarks. Conduct thorough research to avoid potential legal issues.

Design an eye-catching cover: A visually appealing book cover is crucial for attracting potential buyers. You can ask ChatGPT for a list of bestsellers in your book category, so you can study their cover designs and gain inspiration. Then, you can create your cover using design tools like Canva or Adobe Spark, leverage AI-based design tools like Midjourney, or hire a professional graphic designer through platforms like Fiverr, Upwork, or 99designs.

Next, you want to properly format your book: Properly format your book according to Amazon KDP's guidelines. Use Amazon's free Kindle Create tool or other software like Atticus, Vellum or Scrivener to format your manuscript for Kindle e-books and print books.

Create an Amazon KDP account: Sign up for a free Amazon KDP account. Fill in your personal and payment information, as well as tax information.

Publish your book on Amazon KDP: Upload your manuscript and cover, enter relevant information such as book title, author name, description, and keywords. (Ask ChatGPT to provide you with profitable, and highly-searched keywords related to your book)

Choose the appropriate categories and set your book's pricing. For e-books, you can enroll in KDP Select to take advantage of promotional tools like Kindle Countdown Deals and Free Book Promotions.

Optimize your book listing: Ask ChatGPT to craft a compelling book description for your book that highlights the benefits and value of your book. Use relevant keywords in the title, subtitle, and description to improve search visibility. Also, encourage readers to leave reviews to enhance your book's credibility.

Develop a marketing strategy: You want to brand yourself. So create an Author page on Amazon KDP, and get ChatGPT to help you fill out your bio, and come up with usernames for Instagram, Tiktok and Facebook.

Next, ask ChatGPT about book launch strategies and how to use free promotions to generate buzz and increase sales. Create a marketing strategy for your latest release. Ask ChatGPT how to promote your book through various marketing channels, including social media, email marketing, and content marketing. Ask it to help you leverage Amazon Advertising. And ask ChatGPT to help you collaborate with influencers and other authors in your niche to increase visibility and sales. For example, it can generate messages you can send to influencers for a collaboration, or to offer them a free book in exchange for an honest review, etc.

Eventually, you will want to create an author website as well. But take things one step at a time in the beginning.

Track sales and adjust your strategy: Monitor your book's sales performance using the KDP dashboard. Analyze the data to identify trends and adjust your marketing strategy, pricing, and keywords to optimize your book's performance.

By following these steps, you'll be well on your way to self-publishing with Amazon KDP and successfully selling your books. Remember to stay persistent and continuously refine your strategies to maximize your success.

Chapter 6

Networking and Collaboration with ChatGPT to Build Wealth

In this chapter, we'll explore how leveraging ChatGPT can not only enhance your personal and professional growth, but also help you build wealth through networking and collaboration. We'll focus on some key areas where ChatGPT can help you connect with others and create opportunities: finding mentors and business partners, engaging with the ChatGPT community, crowdsourcing ideas and resources.

Finding Mentors and Business Partners

ChatGPT can play a pivotal role in helping you find like-minded individuals who can support and guide you on your journey to becoming a millionaire. After all, it is said that we become like the five people closest to us. And, if you surround yourself with successful people, or with millionaires, you will become the next successful millionaire in the group.

1. Identifying potential mentors: ChatGPT can help you create a list of potential mentors based on your interests, goals, and industry, as well as provide guidance on how to approach and establish relationships with them.

2. Business partner search: ChatGPT can assist you in finding potential business partners by generating profiles of ideal candidates, suggesting networking opportunities, or even helping you draft partnership proposals.

3. Building relationships: ChatGPT can offer tips on fostering strong professional relationships, effective communication, and maintaining long-term connections with mentors and partners.

4. Collaborative projects: ChatGPT can help you brainstorm ideas for collaborative projects, facilitating partnerships that can contribute to your financial success.

Engaging with the ChatGPT Community

Connecting with the ChatGPT community can provide valuable insights, resources, and opportunities:

1. Online forums and social media: ChatGPT group can direct you to online forums, social media groups, and other platforms where ChatGPT enthusiasts gather to share ideas, insights, and experiences.

2. Collaboration: ChatGPT can help you to find people in your field who you can share your audiences with so that you both grow. It can do this if you ask it where to find others in your niche who are doing the same or similar things as you, ex. On LinkedIn, podcasts, Youtube, Tiktok, etc. You could then ask ChatGPT to construct an email to reach out to them.

3. Networking events: ChatGPT can inform you about upcoming networking events in your

niche, conferences, or meetups related to your niche, that provide opportunities to connect with others in the field.

4. Learning from peers: ChatGPT can help you identify and connect with individuals who share similar interests or goals, fostering a mutually beneficial exchange of knowledge and resources.

5. Building your personal brand: ChatGPT can guide you in creating a compelling online presence, showcasing your skills and expertise within your community and beyond.

Crowdsourcing Ideas and Resources

ChatGPT can facilitate the process of tapping into the collective wisdom of the crowd to generate ideas and resources for your wealth-building journey:

1. Idea generation: ChatGPT can help you create and manage online surveys, polls, or brainstorming sessions to gather ideas and feedback from a diverse audience.

2. Crowdfunding campaigns: ChatGPT can assist you in designing and launching crowdfunding

campaigns to raise capital for your business or investment projects.

3. Collaborative problem-solving: ChatGPT can help you engage with online communities to find innovative solutions to challenges you face in your wealth-building endeavors.

4. Skill and resource sharing: ChatGPT can facilitate connections with individuals willing to share their skills, expertise, or resources, fostering a mutually supportive network.

Prompts to help you network and collaborate to build wealth and grow your brand

1. How can I find and connect with like-minded individuals in [niche/area of interest] who are interested in networking and collaborating?

2. What are the best online platforms or communities for discovering potential collaborators and networking opportunities within the [niche/area of interest]?

3. What strategies can I use to effectively network with professionals in [niche/area of interest] to

explore collaboration and wealth-building opportunities?

4. Can you recommend any events, conferences, or meetups specifically tailored to [niche/area of interest] where I can network with others and discuss potential collaboration for wealth creation?"

5. What are some examples of successful collaborations in [niche/area of interest] that have led to wealth generation, and how can I learn from their experiences?

6. How can I leverage social media platforms like LinkedIn, Tiktok, Twitter, Instagram or Facebook to find and connect with individuals interested in networking and collaboration within [niche/area of interest] for wealth building purposes?

Chapter 7

How to Use ChatGPT Safely to Protect Your Data

When it comes to leveraging ChatGPT, many discussions go on about safety and security. But one crucial aspect that often gets overlooked is plagiarism. You never want to get in trouble with the law, so it's important you double-check the responses from ChatGPT to ensure there are no copyrighted materials. This is especially true for works you aim to publish or benefit from financially.

Two solid choices for this include:

Originality AI — a platform tailored for content creators that specializes in plagiarism checks and AI detection

Grammarly — the tried and true favorite of writers everywhere.

Be careful with what you upload

ChatGPT is open source, and it takes in content from its users as well as giving out content to them- so it's important you don't put in any personal information or full documents that you created. That means don't upload whole stories, business plans, etc., as someone else may use them without your permission in the future, which could breach copyright and expose your ideas. So use ChatGPT for data, but don't be a source for it.

Saving your work

Protecting your data does not only entail not submitting your original works and private data into the chatbox. It also means protecting any information ChatGPT generates for you. You will find your previous "chats" with the AI model are located on the left sidebar of the program. However, it is highly advised that you back up your important chatlogs onto your computer or into a program like Google Docs. This is because the platform is not perfect yet, and people have experienced entire chat logs permanently disappearing, which has cost them tons of information and work. You want to avoid things like this that waste your time and can cause frustration.

Here are two methods of preserving your chat dialogue

1. Copy the message chain from the chatbox and paste it into a word processing program. Afterwards, save it on your computer, upload to Google Docs, email, or cloud storage.

2. Take pictures of the chatlog on either a laptop or cellphone. This will create an image file which can be stored on your computer, shared through Google Docs, emailed, or uploaded to the cloud.

Chapter 8

Conclusion - Building Wealth with ChatGPT

In this final chapter, we'll reflect on the exciting possibilities that ChatGPT presents for wealth creation and financial success. We'll discuss the future of wealth creation with AI, the importance of embracing change and innovation, achieving financial freedom through ChatGPT, and share inspiring success stories and case studies.

The future of wealth creation with AI

As AI technology continues to advance, the potential for wealth creation will grow exponentially. ChatGPT and other AI-driven tools will play a central role in shaping the future of various industries, offering new opportunities for innovation and value creation.

Disrupting traditional industries: It's been said that everything we know right now, has a half-life of 2 years. Meaning that technological advancements are happening so quickly that, what you know at this moment, will not matter much, within the next 2 years. This can be a frightening concept for most people. It's one of the unfortunate things about technology. New technology has historically always disrupted industries.

ChatGPT is no different. It is already transforming existing industries. But this does not have to be a bad thing. In fact, AI technology can create new markets and revenue streams for millions of people, and offer us all endless opportunities for wealth creation!

1. Democratizing access to information and resources: AI will make it easier for individuals to access knowledge, tools, and networks, leveling the playing field and enabling more people to build wealth.

2. Personalizing financial strategies: AI will provide tailored financial advice and strategies, helping individuals optimize their wealth-building ef-

forts.

3. AI-driven entrepreneurship: AI will empower entrepreneurs to create innovative products and services, driving economic growth and wealth generation.

Embracing change and innovation

It's human nature to fear what is new and different, but it will be the early adopters of change that will reap the most benefits. People who are quick to learn about new opportunities, applications, and innovations in the world of technology and business tend to make more money and have more authority in their field on average, compared to the rest of the general public who adopts a technology later on. Gary Vee, the American entrepreneur, speaker, internet personality and social media guru, worth $200 million, constantly tells aspiring entrepreneurs, who aim to be wealthy, to always try to be one of the first people to adopt any new social media application. So many businesses have blown up thanks to social media, including my own. You never know when you, or one of your products or posts will go viral, which is why it's so important to stay relevant. Sign up to new apps, use the latest AI models.

Go where the people are and advertise your brand. Be consistent, and you may just be surprised one day.

Achieving financial freedom through ChatGPT

ChatGPT has the potential to transform your wealth-building journey and help you become a millionaire. It can help you manage your finances and automate work, therefore saving you precious time and energy to focus on what really matters. Use it to reap the benefits of data-driven insights and recommendations, and let it empower you to make smart financial decisions. By leveraging ChatGPT's capabilities, you can uncover new investment and business opportunities, learn how to create profitable online businesses and accelerate your goal to wealth and financial freedom! And lastly, take advantage and use ChatGPT as a catalyst for personal growth and self-improvement, equipping you with the knowledge, skills, and confidence to achieve your financial goals.

Harnessing the power of ChatGPT can open up a world of possibilities when it comes to financial success. In order to make the most of ChatGPT's capabilities, you need to stay informed, adaptive and open to innovation as AI continue to evolve quickly.

ChatGPT provides an unrivaled wealth of opportunities for individuals looking to achieve financial freedom.

From investment optimization and budgeting, to creating new products and services, ChatGPT has all the potential to revolutionize how we approach wealth creation.

As you strive for achieving your millionaire dreams, remember that determination, resilience and dedicating yourself to learning are invaluable assets. With ChatGPT as your ally, you have access to resources, guidance and support to turn your ambitions into reality.

So don't hesitate to seize the moment and embrace the future of AI-driven wealth creation with ChatGPT! Let AI take you by the hand, and become your personal financial mentor on your journey to reaching 7-figures. With ambition, savviness and creativity, there is no limit to what you can do! You can absolutely become the first millionaire in your family. In fact, mark my words, you will be.

Chapter 9

---

30 Best Prompts to Ask ChatGPT In Order to Create $1 Million As Fast As Possible

These done-for-you prompts are ready to be plugged into ChatGPT. They are meant to get ChatGPT thinking of the best and fastest ways you can earn $1 million quickly, effectively and of course, legally. They are also meant to inspire you to seek out and try several diverse business models you may not have heard of or considered before. Pick one at a time, and stick with it for at least 3

– 6 months in order to see results, and before deciding it "doesn't work."

The idea is to try as many different business models as possible to see which ones you enjoy most, and which ones work for you, as well as which ones don't. You may be wondering, "what if some of these business models don't work for me?" "What if I fail?"

That's the idea! You should try and "fail!" And you should do this a lot. Why? Because the average millionaire has tried 17 different businesses, enterprises, concepts and did not succeed until their 18th try! Basically, in order to become a millionaire you must let go of the *fear of failure*. Reprogram your thinking so that you start to see every "failure" as nothing more than a lesson. Soon, your brain will learn that there is no such thing as "failure." You only win or you learn.

1. "What are the most profitable online business models to pursue for generating $1 million in revenue quickly?"

2. "How can I identify and capitalize on emerging trends or niches in e-commerce?"

3. "What are the best strategies for creating a high-converting sales funnel to maximize revenue

generation?"

4. "How can I effectively utilize affiliate marketing to boost my income and reach the $1 million goal?"

5. "What are the key principles for designing a successful subscription-based business model?"

6. "How can I leverage content creation and marketing to build a profitable online brand?"

7. "What are the most effective strategies for growing a profitable email marketing list?"

8. "How can I optimize my e-commerce store to increase average order value and overall revenue?"

9. "What are the best practices for pricing digital products or services to maximize profitability?"

10. "How can I use influencer marketing to quickly scale my online business and reach a wider audience?"

11. "What are the most successful tactics for promoting and selling high-ticket items or services online?"

12. "How can I create a profitable online course or educational platform in a high-demand niche?"

13. "What are the best methods for building a strong personal brand that can generate significant income?"

14. "How can I develop a highly effective content marketing strategy to drive traffic and increase sales?"

15. "What are the top strategies for outsourcing and automating tasks to scale my online business rapidly?"

16. "How can I utilize webinars to sell high-value products or services and generate significant revenue?"

17. "What are the best practices for building and monetizing a large and engaged social media following?"

18. "How can I create and sell a successful software product or SaaS solution to reach my $1 million goal?"

19.

"What are the top methods for driving consistent, high-quality traffic to my online business?"

20. "How can I leverage strategic partnerships and collaborations to rapidly grow my online revenue?"

21. "What are the best techniques for building a highly engaged and responsive online community?"

22. "How can I optimize my website's SEO to drive organic traffic and increase sales?"

23. "What are the most effective strategies for creating and selling digital products, such as eBooks or templates?"

24. "How can I use podcasting as a platform to generate income and reach my $1 million target?"

25. "What are the best methods for successfully launching a new product or service online?"

26. "How can I create a viral marketing campaign to boost awareness and drive sales for my online business?"

27. "What are the top strategies for using YouTube to

generate significant income and grow my online presence?"

28. "How can I develop a profitable membership site or community in my niche?"

29. "What are the most effective techniques for converting website visitors into paying customers?"

30. "How can I leverage customer testimonials and social proof to increase trust and boost sales for my online business?"

Chapter 10

Resources

With brand new AI tools popping up nearly everyday, there are now so many tools that can help you create a profitable online business.

Understandably, you may have no idea where to begin, and that is why we wanted to provide you with what may very well be the only website you will need to stay up-to-date with the latest, most useful and amazing AI tools. Because, after all, ChatGPT is really just the beginning.

Check out this website: Future Tools - Find The Exact AI Tool For Your Needs

And check out over 1000 tools that can help you create profitable products, marketing strategies, content, manage your time better, create art, summarize and read books, and so much more!

Author's Notes

T hank you for reading. If you enjoyed the story or have encouraging or constructive comments please leave a review! It also helps more readers discover my work, so thank you in advance!